Mohammad Emdadul Haque

Earthquake Resistant Design, Analysis and Construction to Registration

Mohammad Emdadul Haque

Earthquake Resistant Design, Analysis and Construction to Registration

Noor Publishing

Imprint
Any brand names and product names mentioned in this book are subject to trademark, brand or patent protection and are trademarks or registered trademarks of their respective holders. The use of brand names, product names, common names, trade names, product descriptions etc. even without a particular marking in this work is in no way to be construed to mean that such names may be regarded as unrestricted in respect of trademark and brand protection legislation and could thus be used by anyone.

Cover image: www.ingimage.com

Publisher:
Noor Publishing
is a trademark of
Dodo Books Indian Ocean Ltd. and OmniScriptum S.R.L publishing group

120 High Road, East Finchley, London, N2 9ED, United Kingdom
Str. Armeneasca 28/1, office 1, Chisinau MD-2012, Republic of Moldova, Europe
Managing Directors: Ieva Konstantinova, Victoria Ursu
info@omniscriptum.com

Printed at: see last page
ISBN: 978-620-8-87105-5

BISMILLAHIR RAHMANIR RAHIM

Earthquake Resistant Design, Analysis and Construction to Registration

First Edition

Mohammad Emdadul Haque

Brief Contents

BISMILLAHIR RAHMANIR RAHIM

Chapter-1

Earthquake Resistant Design and Liquefaction Factor Investigation Being Bangladesh in High Probable Seismic Region

Introduction

1.1 General

"Did We not make the earth habitable? And the mountains as pegs? (78:6-7)"The earth is formed of several layers that have very different physical and chemical properties. The outer layer, called the 'Lithosphere', is made up of the crust and the upper most part of the mantle. The Lithosphere averages about 70 kilometers in thickness and consists of about a dozen large, irregularly shaped plates. These plates slide over, under and past each other on top of the partly molten inner layer because of gradients set up by thermal isostatic processes. Thus, warping and connective movements begin and the plates move, carrying the continents and oceans with them. [1]. Most earthquakes occur at the boundaries where the plates meet. Figure 1.1 shows Global Tectonic Plate. There are three main types of plate boundaries i.e. Spreading Zones or Divergent Boundaries, Transform faults or Transform Boundaries and Convergent Boundaries. These boundaries are subjected to different inner plate stresses such as- Extensional stress, Transform stress and Compressional stress. Earthquakes can also occur within plates, although plate boundary earthquakes are much more common.

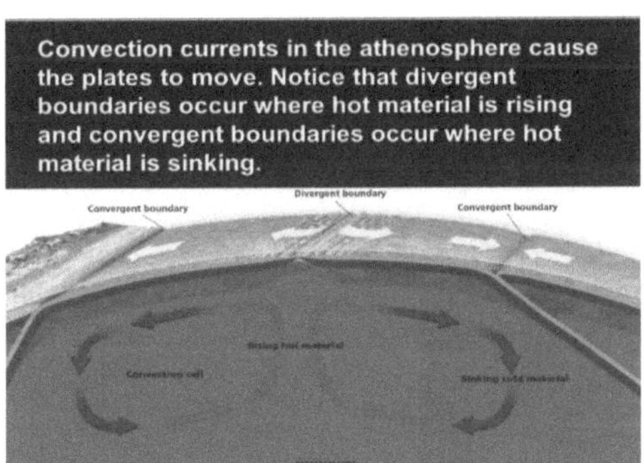

The most important seismic hazards include ground shaking, structural failure, soil liquefaction, landslides, and lifeline hazards. The ultimate goal of the earthquake engineer is to improve earthquake-resistant design, so that structures is going to be strong enough to sustain design-basis earthquakes. Further understanding of the structural dynamic responses during strong earthquakes requires significant collaboration involving engineering geologists, seismologists, geo-technical engineers and structural engineers.

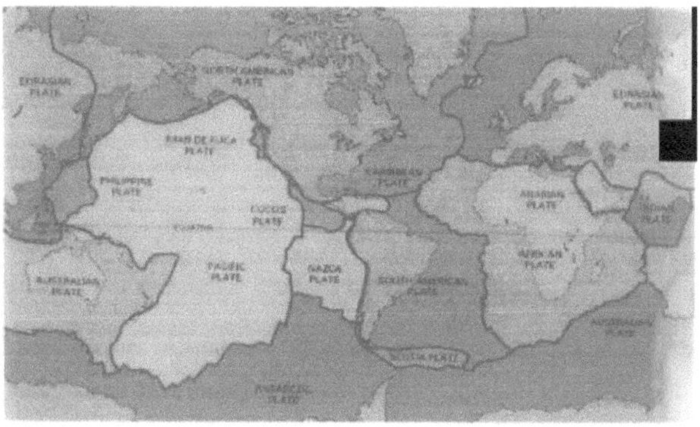

Figure 1.1: Global Tectonic Plates

Bangladesh is situated in the delta region of the Bay of Bengal. The main active fault in that region, the Andaman fault, is becoming active day by day. According to seismologists, the slip rate of that fault is 63 mm/year. This fault crossed Bangladesh and met Himalayan fault through India. The minor earthquakes of magnitude ranging from 3-4 (Richter Scale) are the common phenomena in Bangladesh. There are records of strong motion earthquakes in Bangladesh. These earthquakes which occurred in Indian territorial boarder, were also felt in Bangladesh. In order to understand the seismicity of the country, Bangladesh is divided into three zones. Roughly it may be stated that the north - eastern part of Bangladesh is more prone to earthquake shocks than south - eastern part. This does not mean in any way that strong earthquakes are ruled out in the south - eastern part and south - western part. It is estimated that almost the total area of Bangladesh is in the active seismic zone.

It may be noted that the 1897 great Indian earthquake is considered to be one of the strongest earthquake in the world originated at a epicentral distance of only 230 km from Dhaka, while the earthquake affected almost whole of Bangladesh. The damages were very severe particularly in Sylhet, Rangpur and Mymensingh. In the city of Dhaka most of the brick masonry buildings either collapsed or were severely damaged. However due to increasing number of buildings, bridges and industrial being built during the last two decades, proper assessment of hazard due to earthquakes is essential. Under estimation of the hazard may lead to unsafe structures, whereas overestimation will lead to unnecessary increase in investment.

1.2 Code Provisions

A building code is a set of legal requirements intended to ensure that the location, design and construction of the building, if subjected to natural or manmade destructive forces, it will present no significant threat to life, health or welfare of its occupants or the general public. Seismic Codes are based on knowledge, derived from experience, laboratory testing and theoretical analysis. To ensure the protection of human life a building must survive earthquakes without collapsing. Provisions of the Bangladesh National Building Code account for the probable seismic hazards in any given region. A fundamental aspect of seismic design is that yielding is allowed to accommodate seismic loading as long as such yielding does not impair the vertical load capacity of the structure. It would be uneconomical to design building so that it remains elastic at all levels of earthquake ground motions.

1.3 Objectives of Study

The study deals with the earthquake resisting reinforced concrete structures and rehabilitation of the structures which are not designed for earthquake or which are

already damaged by earthquake. The main objective of this work are to illustrate the structural engineering problems and its solutions in attaining efficient earthquake resistant structures. Thus the main objectives of the thesis are:

1. To study the structural performance during earthquake.
2. To study the general concept of seismic design.

3. To investigate site condition for determining factor of safety against earthquake.

4. To study the principles of earthquake resistant design of reinforced concrete structures.

5. To determine earthquake load, designing and reinforcement detailing of earthquake resistant structures.

1.4 Scope of the study.

To achieve this objective of seismic resistant design and Rehabilitation of structures, some topics are discussed in this thesis. In Chapter 2 earthquake in Bangladesh, structural performance during earthquake and general concept of seismic design are discussed.

The site condition of soil is analyzed to determine the factor of safety against earthquake in Chapter 3. The principles of earthquake resistant design and analysis are presented in Chapter 4. In Chapter 5 a sample building is analyzed and designed for earthquake load.

Analysis of Site Condition

1.5 Pre- requisite Conditions For Liquefaction to Occur

- Soil deposit: Granular soil such as silty sand of or sandy silts with or without fine content.
- Soil should be **saturated,**
- Load application should be repeated or cyclic in nature so that pore pressure can be built up gradually.
- Loading rate should be high enough to establish undrained conditions in the cohesionless soil mass.

1.6 Representation of cyclic Strength

$$\text{Cyclic Strength} = \frac{(\text{Maximum axial stress-Horizontal pressure})}{2*(\text{Effective all round stress})}$$

$$= \text{Deviator stress}/2\ (\text{Effective all round stress})$$

1.7 Factors affecting cyclic strength/stress ratio of soil deposit

- **Density of soil deposit** : higher density, higher strength.
- **Fine content**: Presence of fine content (i.e. material passing through# 200 sieve) increases cyclic strength
- **Soil plasticity** : does not have any effect if the plasticity index Ip < 1 0; but if not, the cyclic strength increases slightly with the increase of Ip.
- **Depositional history** : Freshly deposited or dry deposited (i.e. deposition from zero height) sands exhibit lower cyclic strength than air pluriated deposit or a mass reconstituted by moist tamping.
- **Ageing & inherent fabric**: The cyclic resistance of undisturbed samples from the in situ deposit is about twice as great as the cy clic resistance of the sample from the newly deposited sand fill.

1.8 Sample Problem for Analyzing of Site Condition to Determine Factor of Safety Against Earthquake

The sample problem is taken of a site . The data is taken for soil deposit at the Civil Building Site of BUET and the data taken are assumed data.

a. By means of SPT , penetration resistance N is obtained , together with the fines content, Fe or mean diameter D50 for the soils in question throughout the the depth at a given site

b. If necessary , the plasticity index Ip of the fines fraction is determined .

c. The percentage of particles passing through # 200 sieve is estimated .

Following are the data of soil deposit at the Civil building site of BUET, Dhaka for depths up to 7.5m below ground level (GL):

i. SPT Nvalues at depths 1.5, 3.0, 4.5, 6.0 and 7.5 are 1, 2, 3, 5 and 9, respectively.

ii. Ysat of soils at depths 1.5, 3.0, 4.5, 6.0 and 7.5-m are 17.0, 17.5, 17.0, 17.8 and 18.3 kN/m3 respectively.

iii. Ydry of soils at depths 15.5, 3.0, 4.5, 6.0 and 7.5-m are 13.0, 13.5, 13.3 and 14.3 kN/m3, respectively.

iv. Water table is at 1.5m below the GL and Ywater = 9.8 kN/m3

v. Plasticity index = 17% at all depths

Estimation of the liquefaction of the soil deposit at the depth of 6m below GL if the grain size distribution of soil at the particular depth shows: D5o = 0.21mm and particles passing# 200 sieve = 18%. According to BNBC (1993), Dhaka lies in Zone II and for this particular zone, the design value of the ground acceleration due to the anticipated earthquake is 0.15g, where g is the acceleration due to gravity.

Solution:

Plasticity Index, lp = 17% (For All Depth)

Ywater = 9.8 KN/m3

Dso = 0.21mm

Particle Passing # 200 = 18%

Ground acceleration due to anticipated Earthquake = 0.15g

Liquefaction potential at depth of 6m below Ground Level = ?

Various data and calculations are shown in Table 3.1

Table 3.1
Calculation At Various Depth

Depth (m)	N	γ_{sat}	γ_{dry}	Total Stress	Effective Stress	Total Stress/ Effective Stress	$C_N = (1.7)/$ (Effective Stress+1.7)	$N'_1 = N*C_N$	N_1
1.5	1	17	13.0	19.5	19.5	1	0.895	0.895	
3	2	17.5	13.5	45.75	31.05	1.473	0.845	1.69	
4.5	3	17	13.5	71.25	41.85	1.703	0.8	2.4	
6	5	17.8	13.3	97.95	53.85	1.82	0.76	3.8	10.3
7.5	9	18.3	14.3	125.4	66.6	1.88	0.72	6.43	

$\triangle N_1 = 0.5\,(F_c-5)$ for $5 \le F_c \le 20$

$= 7.5$ for $Fc > 20$

$\triangle N_1 = 0.5\,(18-5) = 6.5$

Design N1 $= N'_1 + \triangle N_1$

$= 3.8+6.5 = 10.3$

Using Japanese code:

There are Chinese and Japanese Code . We are using here the Japanese Code .
For $D_{50} = 0.21$mm

Cyclic Strength / Resistance $= 0.0676\,\sqrt{N_1} + 0.225\,\mathrm{LOG}_{10}(\frac{0.35}{D_{50}})$

$= 0.0676\,\sqrt{10.3} + .225\,\mathrm{LOG}_{10}(\frac{0.35}{0.21})$

$= 0.267$

PLASTICITY INDEX

PI = 17%

For $I_P > 10\%$, R $= 1/45\,(I_P-10) + 1$

So, R $= 1/45\,(17-10) + 1$

$= 1.16$

The Design Cyclic Strength $= 0.267 \times 1.16$

$= 0.3085$

$r_d = 1 - .015Z$

$= 1 - .015\times 6 = 0.91$

$a_{max} = 0.15g$ (Dhaka Zone)

Distributing Stress $= \frac{a_{max}}{g} \cdot r_d \cdot$ (Total Stress / Effective Stress)

$= \frac{0.15g}{g}\ 0.91 \times 1.82$

$= 0.2484$

Factor of Safety = Cyclic Strength/ Disturbing Strength = $\frac{0.3085}{0.2484}$ > 1 ,The above investigation shows that Soil will not be liquefied
In other words site Condition is Safe Against Earthquake.

1.9 Code Provisions for Joint Design

- Beam longitudinal reinforcement terminated in a column shall be extended to the far face of the confined column core and anchored in tension according to the requirement given below:

- Transverse reinforcement used in column must be continued through the joint . The reinforcement may be reduced by one -half within the depth at the shallowest framing member and the spacing of spirals or hoops may be increased to 6 inch, if beams or girders frame into all four sides of the joint and flexural member core at least 3/4 of column width.

- Joint reinforcement detailing as per ACI-99 code is presented in Figure 7.1 a,b,c.

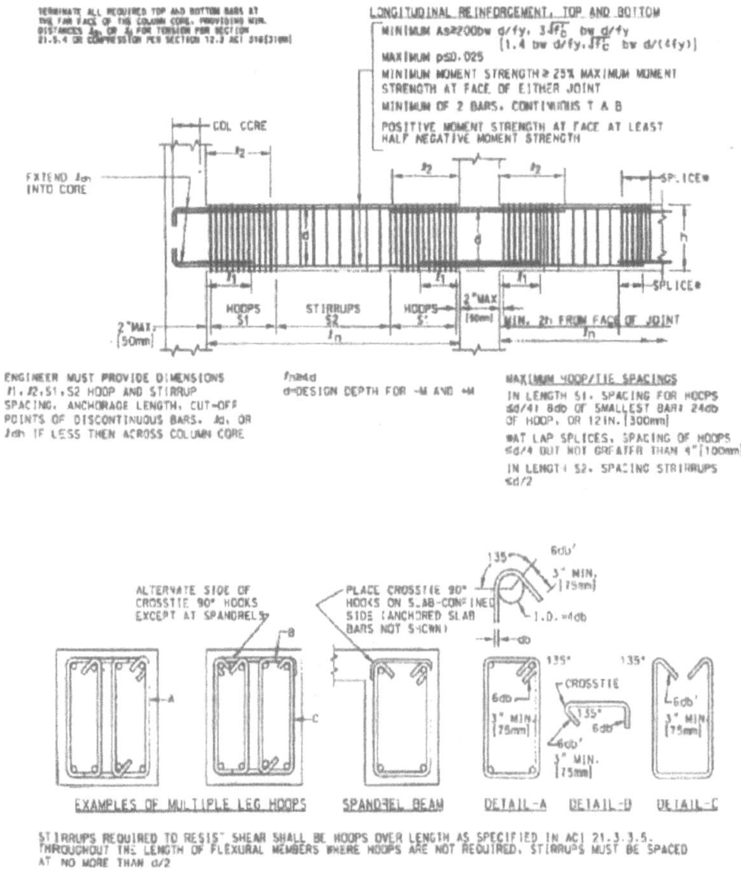

Fig. 5-Typical seismic-resistant details: Flexural Members.

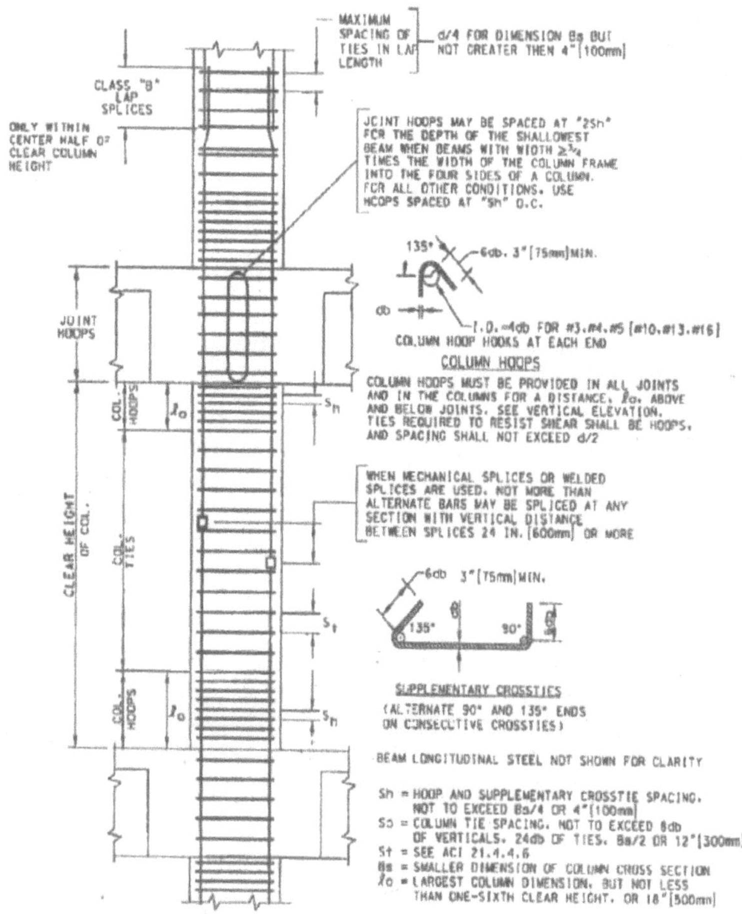

Fig. 6-Typical Seismic Reinforcement Details: Columns.

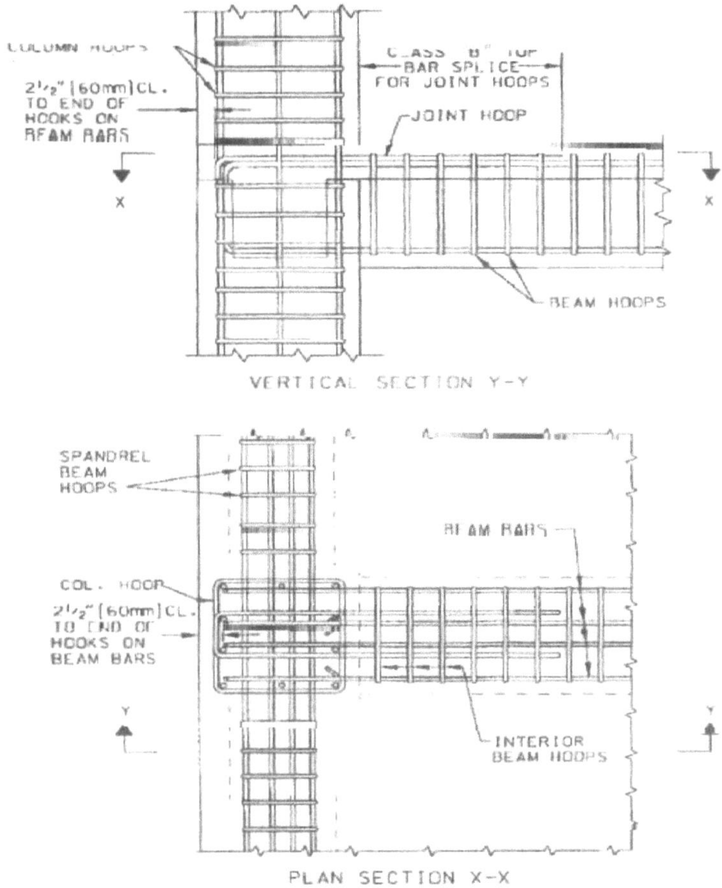

Fig. 7(a)-Typical seismic-resistant joint details-Case 1: For regions of high seismic risk. Interior and spandrel beams narrower than column.

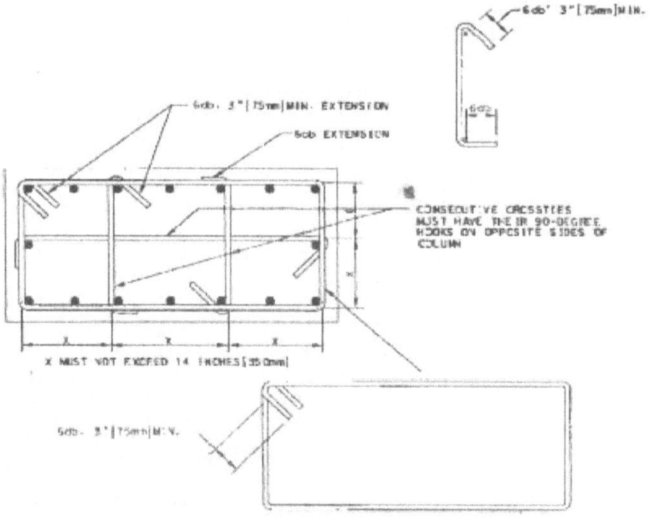

Fig. 17—Typical seismic resistant details: transverse reinforcement in columns.

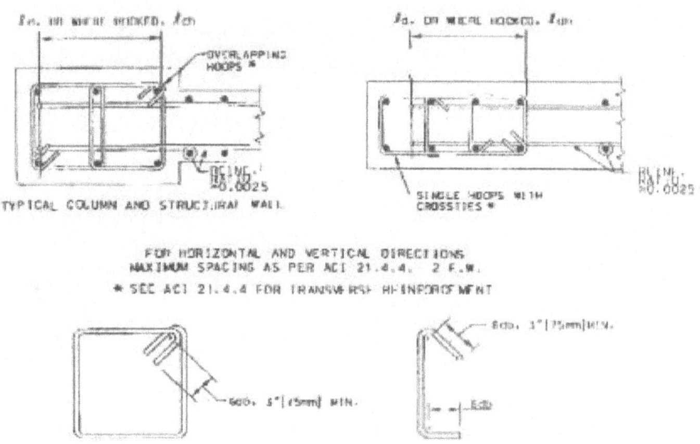

Fig. 18-Typical seismic-resistant details: boundary members.

Analysis of Sample Building

A sample building is taken to design as Earthquake resistant Structure. The dimensions of the building is shown in plan. There are four loads which are considered here .Those are Dead load , Live load , Wind load and Earthquake load .Analysis is done by four combinations, those are:

1. Case 1 : (1.4 DL+ 1.7 LL)
2. Case 2: (1.4 DL+ 1.7 LL + 1.7 WL)0.75
3. Case 3: (1.4 DL + 1.7 LL-1.7WL)0.75
4. Case 4: (1.4 DL+ 1.7 LL+ 1.87 EQ)0.75

In sample calculation of Earthquake load and Wind Load ,forces at each Floor is found out . The calculation has followed BNBC Code . Then from maximum moment at midspan among all the flexural members the Beam is designed . Maximum Shear force is found out among all the edges , and from that Column is designed along with shear reinforcement .

Analysis

Longer direction

Total 168 ft@ 24 ft per bay.

Figure 5 .1: Typical Floor Plan of Example Building

Design Data :

Live Load :50 psf

Superimposed dead load: 42.5 psf

For concrete: fc' =4 ksi

For steel : fy= 40 ksi

Basic wind speed = 210 km/hr for Dhaka ,Zone 2

Earthquake Load Calculation (Longer Side)

Seismic Zone Coefficient, Z= 0.15 , Since Dhaka is in Zone 2

Structural Importance Coefficient =1 , for standard occupancy structures

Response Modification Coefficient for Structural Systems, R=12 (for special moment resisting frame where it may be steel or concrete)

S=S3 = A soil profile 21 metres or more in depth and containing more than 6 metres of soft to medium stiff clay but not more than 12 metres of soft clay = 1.5

Ct = 0.073 for reinforced concrete moment resisting frames, and ecentric braced steel frames

Zone- 2, Z = 0.15

I = I

SMRF, Concrete

R = 12

T = Ct(hn)3i4

=0.95> 0. 7sec

S = 1.5

$$C=\frac{1.25\ S}{T^{2/3}}$$

= 1.94<2.75; OK

W=total DL

=(2x168x10)

$V = \frac{ZIC}{R} . W$

$= \frac{(O.15)(l)(1.94)}{(12)} = (3360) = 81.48K$

F= 0.07T.V

= 0.07 (0.95) (81.48)

= 5.42

0.25V=20.37

Ft≤0.25V; OK

Wx= DL of each floor = 2x 168 = 336 k

€ Wihi = (336) (10+20+30+40+50+60+70+80+90+ 100)

= (336) (550)

= 184800

(V-Ft) Wx hx

$F_x = \frac{(V-Ft)Wx\ hx}{Wihi}$

$= \frac{(81.48-5.42)336)hx}{(1848000)} = 0.14\ h_x$

Forces at Each Floor (Longer Side)

Storey	h_x	F_x (K)
1	10	1.4
2	20	2.8
3	30	4.2
4	40	5.6
5	50	7
6	60	8.4
7	70	9.8
8	80	11.2
9	90	12.6
10	100	14

Earthquake Load Calculation (Shorter Side)

$W = 2\times60\times10 = 1200K$

$V = \frac{ZIC}{R} W = \frac{(0.15)(1)(1.94)}{(12)} (1200) = 29.1 K$

$F_t = 0.07T.V = 0.07(0.95)(29.1) = 1.935$

$W_x = 2\times60 = 120K$

$\Sigma Wihi = 120\times550 = 66000$

$F_x = \frac{(29.1-1.93)(120)hx}{(66000)}$

$= 0.05hx$

Forces at Each Floor(Shorter Side)

Storey	hx	Fx (K)
1	10	0.5
2	20	1
3	30	1.5
4	40	2
5	50	2.5
6	60	3
7	70	3.5
8	80	4
9	90	4.5
10	100	5

Wind Load Calculation (Longer Direction)

$L/_B = \frac{164}{60} = 2.8$

External Pressure Coefficient

$C_{pe} = 0.8$ (for wind ward)

$C_{pe} = 0.26$ (for Lee ward)

Structural Importance coefficient, $C_I = 1$

Check : $\frac{h}{B} = \frac{100}{60} = 1.67 < 5$ (OK)

Designed speed, $V_B = 210$ Km/hr

$q_Z = C_e \times C_I \times C_Z \times 210^2 \times 20.9$

$= 47.2\times10^{-6}\times1\times C_z\times210^2\times20.9$

$q_z = 43.5\, C_z$

Gust response factor,

$C_g = 1.307$

Wind ward side

$P_z = C_g\times C_{pe}\times q_z$

$= 1.307\times 0.8\times43.5\, C_z$

$P_z = 45.5\, C_z$ psf

Leeward side

$P_z = C_g\times C_p\times q_z$

$= 1.307\times0.26\times43.5\, C_z$

$P_z = 15.5\, C_z$

Wind Load Calculation at X Directon

Height (ft)	Combined ht &Exposure Coefficient, Cz	Pz (Psf) Wind ward	Pz (Psf) Leeward	Total Pressure (Psf)
10	.368	16.74	13.3	30.04
20	0.416	18.93	13.3	32.23
30	0.497	22.6	13.3	35.9
40	0.565	25.71	13.3	39.01
50	0.624	28.4	13.3	41.7
60	.677	30.8	13.3	44.1
70	0.7295	33.2	13.3	46.5
80	0.7743	35.23	13.3	48.53
90	0.8158	37.1	13.3	50.4
100	0.856	38.95	13.3	52.25

Wind Load Calculation (Shorter Side)

$L/_B = \frac{60}{168} = 0.36$

External Pressure Coefficient

$C_{pe} = 0.8$ (for wind word)

$C_{pe} = 0.55$ (for wind word)

Structural Importance Coefficient, $C_I=1$

Check , $\frac{h}{B} = \frac{100}{168} = 0.6 < 5$ (OK)

$q_z = C_C = \times C_I \times C_Z \times 210^2 \times 20.9$

$= 4.72 \times 10^{-6} \times 1 \times CZ \times 40^2 \times 20.9$

$q_z = 43.5\ C_z$

Gust respose factor,

$C_g = 1.307$

Wind ward side

$P_z = C_g \times C_{pe} \times q_z$

$= 1.307 \times 0.8 \times 43.5\ C_z$

$P_z = 45.5\ C_z$ psf

Leeward side

$P_z = C_g \times C_{pe} \times q_z$

$= 1.307 \times 0.55 \times 43.5\ C_z$

$P_z = 31.27\ C_z$

Wind Load Calculation at Y Drection

Height (ft)	Combined ht & Exposure Coefficient C_Z	P_Z (Psf) Wind ward	P_Z (Psf) Leeward	Total Pressure (Psf)	Floor Level Load, (Kip)
10	0.368	16.74	26.8	43.5	8.7
20	0.416	18.93	26.8	45.7	9.14
30	0.497	22.6	26.8	49.4	9.88
40	0.565	25.71	26.8	52.5	10.5
50	0.624	28.4	26.8	55.2	11.04
60	0.677	30.8	26.8	57.6	11.52
70	0.7295	33.2	26.8	60	11.9
80	0.7743	35.23	26.8	62	12.4
90	0.8158	37.1	26.8	64	12.8
100	0.856	38.95	26.8	65.75	6.6

Typical Beam Design

Max^m negative moment = -530.94 k –ft

f_c' = 4000 Psi = 4 ksi

f_y = 40 ksi

Let Beam Size 10" ×27"

$$P_b = (.85)^2 \frac{4000}{40000} \frac{87}{87+40}$$

$P = .75P_b = 0.037$

Load on beam = 2 k/ ft

$M_n = \varphi P bd^2 (1 - 0.59 p \ 40/4)$

d = 24.7 = 24.5 "

h = 24.5+2.5=27 " (OK)

$$P_{min} = \frac{200}{fy} = 0.005$$

$As = pbd = 0.037\times10\times24.5 = 9.06 \text{ in}^2$

h = d+2.5 = 27 "

$$As = \frac{M}{\varphi fy\,(d-a/2)} = \frac{531x12}{0.9x40\,(24.5-6/2)} = 8.23 \text{ in}^2$$

$$a = \frac{Asfy}{0.85f'c,b} = \frac{8.25x40}{.85x4x10} = 9.68$$

$$As = \frac{531x12}{0.9x40\,(24.5-9.68/2)} = 9 \text{ in}^2$$

$$a \ \frac{Asfy}{0.85f'c,b} = \frac{9x40}{0.85x4x10} = 10.6$$

$$As = \frac{531x12}{0.x40\,(24.5-10.6/2)} = 9 \text{ in}^2$$

$$a = \frac{Asfy}{0.85f'c,b} = \frac{9.22x40}{0.85x4x10} = 10.85$$

$$As = \frac{531x12}{0.9x40\,(24.5-10.85/2)} = 9.28 \text{ in}^2$$

$$a = \frac{9.28x40}{.85x4x10} = 10.92$$

$As = 9.3 \text{ in}^2$

Use 6#11 Bar

Positive Steel:

Maximum M (+) $= \frac{1}{3}(M-)$

$= \frac{1}{3}(531)$

$=$ 177 K'

As $= \frac{177 \times 12}{(0.9)(40)(24.5 - 5 - 6/2)}$

$= \frac{59}{(24.5 - 6/2)}$

$=$ 2.74 in

a $= \frac{(2.74)(40)}{0.85(4)(10)}$

$= \frac{2.74}{0.85}$

$=$ 3.23

As $=$ 2.6

a $=$ 3.1

As $=$ 2.6 in^2

Use 2 # 11 bars

Then spacing $= 10 - 2 \times 1.5 - 2 \times 1.56$

$=$ 3.88"

Minimum spacing = Minimum bar dia or 1"

$= \frac{10}{8}$" or 1"

$=$ 1"

∴ Bar Spacing = 3.88" > 1"

∴ Ok.

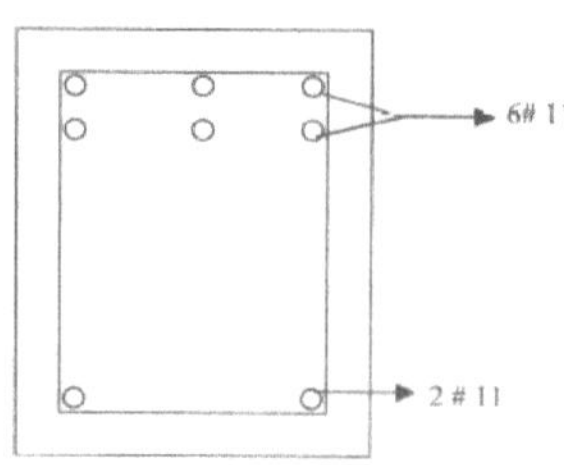

Figure 5.2: Cross Section of Beam

Shear Design of Beam

$V_L =$ } $V_{max} = 52.6$ K
$V_R =$

1) $\varphi V_c = 2\varphi \sqrt{f'c}\ bd$

$= 2\ (0.85)\ \sqrt{4000} \times 10 \times 24.5$

$= 26.34$ k

$\frac{\varphi V_c}{2} = 13.17$ K

$4\varphi \sqrt{f'c}\ bd = 52.68k$

For Part (2) => $V_u = V_{max} - W_d$

$V_s = 52.6 - 6 \times \frac{24.5}{12}$

$= 42.3$ k> φV_c

So web reinforcement in required

$S_{req} = \frac{\varphi A vfyd}{Vu - \varphi Vc} = \frac{0.85x0.22x40x24.5}{(42.3\text{-}21.33)} = 9"$ (using # 3 stirru)

φ $\varphi V_s = Vu - \varphi V_c = 42.3 - 21.33 = 21\ K < (4\varphi \sqrt{f'c}\ bd)$

So, S_{max} :

(1) $\frac{Avfy}{50bw} = \frac{0.22x40x1000}{50x10} = 17.6"$

(2) $^d/_4 = \frac{24.5}{2} = 12.25"/2 = 6.1"$

(3) 24 "

$S_{max} = 6.1$ "

$S = S_{req} = 6.1$ "

use# 3 Closed stirrup @6" 0/0 ,Detailing is shown in fig 5.4 and fig 5.5

This is to be provided within a distance of2h = 2x27 =54"

. ·. Max m allowable stirrup spacing within a distance of 2h (=54 ") is 6"

Beyond 2h,

Smax = d/2 = 12.25" ~12"

:. Use# 3 closed stirrup@ 12"c/c

Typical Colum Design

$f_s = 20$ ksi

$f'_c = 4$ ksi

Let $P_g = 0.025$

Maximum axial load in column = 1000 k

$P = 0.85\ Ag\ [0.25f'_c + P_g \times f_s]$

$1000\ k = 0.85\ A_g\ [0.25 \times 4 + 0.025 \times 20]$

$A_g = 784.31\ in^2$

use 28 "×28 " Tied column

$1000^k = 0.85\ [0.25A_g f'_c + A_g \times P_g \times f_s]$

$= 0.85\ [0.25Ag\ f'_c + A_s\ f_s]$

$= 0.85\ [0.25 \times 28 \times 28 \times 4 + A_s \times 20]$

$As = 19.62\ in^2$

$P_g = \frac{19.62}{28x28} = 0.025$ (Ok)

Provide 20 # 9 bar

Tie bar

1) $16D_b = 16 \times {}^9/_8 = 18$ "

2) $48\ D_b = 48 \times {}^3/_8 = 18$ "

3) Least dimension = 28 "

Tie spacing = # 3 bar @ 18 " $^c/_c$

I. Tie spacing within L distance at column ends -

(i) 1/6 (column clear height)

= 1/6 (9.5')

= 19"

(ii) depth of member

=column depth

= 28"

(iii) previous tie spacing

=18"

Now, L= max^m of the above three

= 28"

Max^m allowable tie spacing over the length L from the joint face is the minimum of

(i) $24d = 24 \times \frac{3}{8}$

= 9"

(ii) 8D =8×1

= 8"

(iii) $h/2 = \frac{28}{2}$

= 14"

∴Tie spacing over L, S = 8"

2. Maxm. allowable tie spacing using # 3 tie outside the distance L

=25

=2 x 8

=16

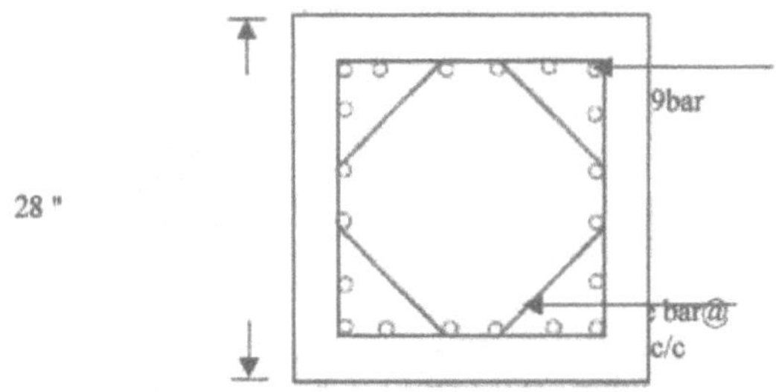

Figure 5.3 : Cross Section of Column

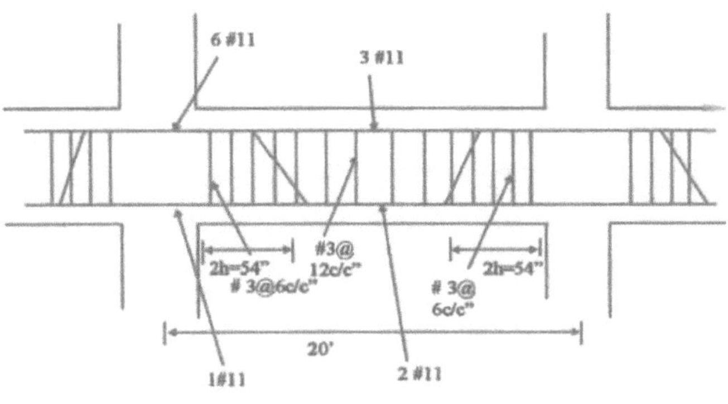

Figure 5 .4 : Beam Detailing

C

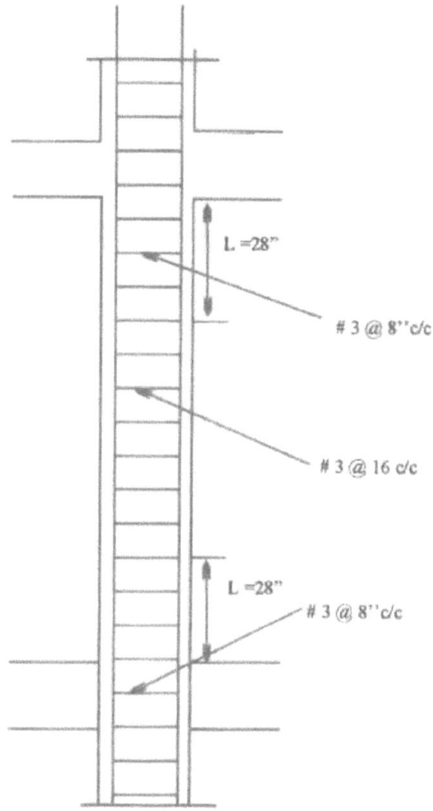

Figure 5 .5 : Column Detailing

Conclusion

1.10 Conclusion

- Floods, cyclones, tornadoes and draughts strike with depressing regularly & intensity but earthquake is added to these damages more severe. The study of earthquake related measure are therefore of immense important in the context of Bangladesh.
- Full analysis is required for both wind and seismic force to find out which force governs for a particular type of building. For the building construction in this study, seismic force governs over wind force
- Soil vibration and Liquefaction is responsible for severe earthquake. Appropriate soil investigation is required before designing a building against earthquake.

BISMILLAHIR RAHMANIR RAHIM

Chapter-2

Earthquake Probability and Inelastic Displacement is Proportional to Building Weight

Introduction

2.1 General

"When the earth is shaken by its (final) earthquake. And the earth discharges its burdens. And man says, what is with it?" (99:1-3). An earthquake is a sudden movement of the earth-crust, caused by the abrupt release of strain, accumulated over a long time. Bangladesh is situated in a seismically active region. Part of the country extended from Sylhet to Chittagong is in the high seismic zone whereas Dhaka is in the moderate seismic zone. Major metropolitan cities of our country are under serious threat because of inadequacies in design and construction of structures.

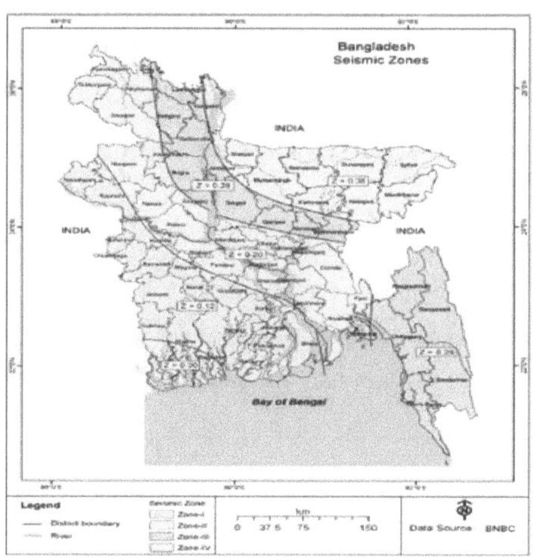

Figure 2.1. Seismic Zoning Map of Bangladesh (BNBC, 2020)

Materials remain elastic at certain level. The ground motion applied on a structure exerts force which becomes nonlinear at certain level. To bring down the ground motion force at inelastic level, Response modification factor is applied. Again, for assessing for total displacement and energy absorbed deflection amplification factor is applied.

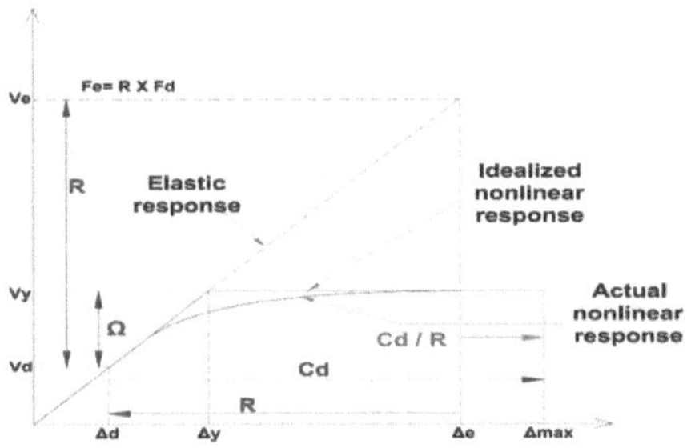

Figure 1.2. Relation Between $\boldsymbol{C_d}$ and $\boldsymbol{R}$ (Astrid et al., 2014)

For estimating the maximum inelastic deflection that might occur during an earthquake, the design deflections computed from an elastic analysis are usually amplified by a deflection amplification factor (C_d). This process is referred to as force-based design method.

2.2 Background and Present State of the Problem

There was no written building code in Bangladesh until 1993. In 1993, Bangladesh National Building Code (BNBC) was published by Housing and Building Research Institute (HBRI) which is commonly known as BNBC. The seismic design provisions of BNBC were based on the UBC (UBC, 1991). For the regular structures, the Code defines a simple method to represent earthquake induced inertia forces by Equivalent Static Force for static analysis. For very tall structure, the Code provisions require Time History Analysis. All these methods detailed in BNBC are force-based methods. As in many other codes, the level of forces prescribed by BNBC for a structure is rather arbitrarily set and aimed at damage control performance objectives i.e., no damage under small earthquake and no collapse under extreme earthquake. This BNBC was gazetted in 2006 as BNBC 2006. Bangladesh National Building Code 2020 (BNBC 2020) is the latest version of building code in Bangladesh. This code is very much similar to ASCE 7-05.

The purpose of this study is to investigate deflection amplification factor of intermediate RC moment frames based on the ratio of maximum inelastic to equivalent static and linear time displacement. Therefore, nonlinear, linear and equivalent static analysis are performed on the 24 models of reinforced concrete moment frames to determine the maximum seismic deflections at all the stories when subjected to the scaled earthquake records.

There are many researches on $C_{d,}$ but as per proposed BNBC no research has been found. In the newly proposed Bangladesh National Building Code (BNBC, 2020), different values for C_d are proposed for different types of RC structures. However, a more detail analysis may be performed to show the variation of C_d due to changes in different RC building parameters like story height (Typical storey height 3.2 meter and bottom storey height 2.13 meter), plan geometry, distribution of mass etc. This would be helpful in understanding the inelastic behaviour of RC building of different configuration designed as per BNBC (2020).

It is to be remembered that the length to breadth ratio of the building in plan not to be higher than 4. In buildings, floors (including the roof) act as horizontal diaphragms that collect and transmit the inertia forces to the vertical structural systems and ensure that those systems act together in resisting the horizontal seismic action.

2.3 Objectives and Scope of the Study

The main objectives of the paper is to determination of elastic and inelastic displacement ratio ($\boldsymbol{C_d}$) and elastic to inelastic storey drift ratio for RC building of different configuration and provide relationship between variations of $\boldsymbol{C_d}$ with different building parameters

2.4 Outline of Methodology

Structural modelling of RC buildings of different configuration to be done in finite element analytical software. 24 intermediate reinforced concrete moment frames to be designed based on BNBC (2020) and ACI 318 (2002) (where needed) provisions. To cover wide range of building geometries, several frame models with two, three and four bays will be selected. Height range of building will vary from 2 to 12 stores, in 1-storey increments except for the frames more than 6 stories, which will have 2-storey increment. The typical bay span and story height will be 5 and 3.2 meters, respectively.

2.5 Existing Analysis and Design Procedure

2.5.1 Equivalent Static Analysis

The evaluation of the seismic loads starts with the calculation of the design base shear which is derived from the design response spectrum, S_a. The building period in the two main horizontal directions to be smaller than both $4T_c$ and 2 seconds and the building does not possess irregularity in elevation.

SC type soil is used in analysis. The parameters in BNBC and Etabs of SC soil has difference. That is why site class F has been used in Etabs for Equivalent static analysis. The parameters of Soil type SC has been taken from Appendix C of BNBC.

2.5.2 Design Base Shear

The seismic design base shear force in a given direction to be determined from the following relation:

$$V = S_a W \qquad (2.1)$$

where,

S_a= Design spectral acceleration (in units of g) corresponding to the building period T.

W = Total seismic weight of the building.

2.5.3 Dynamic Analysis Methods

Spectral acceleration is measured in g that describes the maximum acceleration in an earthquake on an object. Dynamic analysis should be performed for regular buildings with height greater than 40m in Zones 2,3,4 and greater than 90 m in Zone 1. Dynamic analysis may be carried out through response spectrum and linear and nonlinear time history analysis.

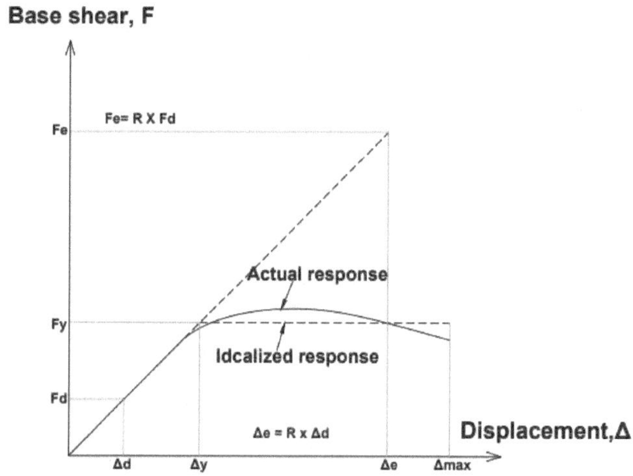

Figure 2.1. General Structural Response under the Effect of Lateral Loading (Elfath, 2019)

Figure shows global inelastic response of a structure under the effect of lateral loading. The actual inelastic response is idealized by a bilinear relation between the base shear and a lateral displacement component of the structure. Δ_{max} represents the maximum displacement demands under inelastic earthquake analysis. The elastic force and displacement demand F_e and Δ_e are related to the design force and displacement demands F_d and Δ_d according to the following relation:

$$\frac{F_e}{F_d} = \frac{\Delta_e}{\Delta_d} = R \qquad (2.15)$$

According to the definition presented by FEMA P695 (2009) and by various researchers such as Uang and Maarouf (1994), the inelastic displacement ratio ρ is calculated according to the following equation:

$$\rho = \frac{DAF}{R} = \frac{(\Delta_{max}/\Delta_d)}{R} = \frac{\Delta_{max}}{\Delta_e} \qquad (2.16)$$

Methodology

2.6 General

Numerical modelling of Reinforced Concrete frames has been presented in this chapter. In this research, total 24 reinforced concrete frame structures have been analysed using finite element method. Firstly, structural analysis of these frames has been performed using equivalent static analysis as per BNBC 2020. Using same geometrical, material and loading data, nonlinear time history analysis has been performed for all cases. There are mainly two methods to evaluate deflection amplification factor. Ratio of nonlinear time history displacement with Equivalent static displacement and ratio of nonlinear time history displacement with linear time history displacement (BNBC, 2020).

2.7 Design Considerations

Structural analysis and design have been performed according to BNBC (2020). Other codes, standards, specifications have been utilized as required in structural design.

A. Structural Geometry Considerations

Initially shape, size, story height and number of stories of the building have been considered as per design requirement and checked as per BNBC (2020). The buildings having bay width of 5 meter and height of 3.2 meter. There are frames of 2,3 and 4 bay. Each bay has 8 frames having 2, 3, 4, 5, 6, 8, 10, 12 story. Typical column location, beam location is shown in the following layout (Figure 3.1).

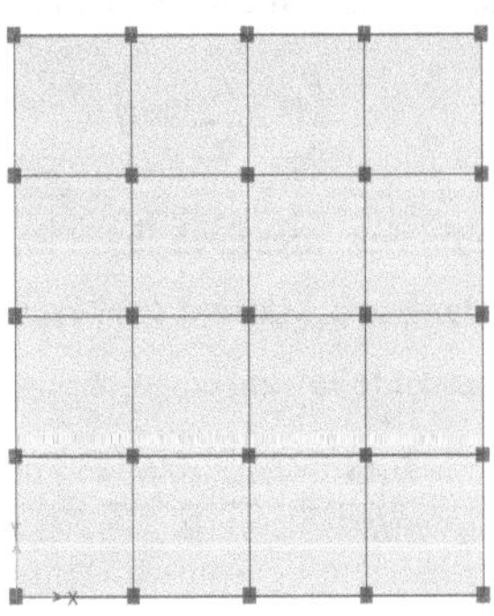

Figure 3.1. Column Layout of Four Bay Twelve Story

2.8 Ground Motion and Analysis Procedure

2.8.1 Peak Ground Acceleration of Imperial Valley Earthquake

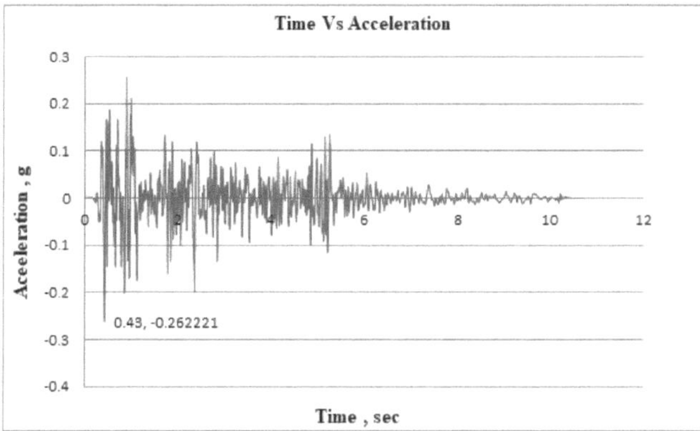

Figure 3.2. Peak Ground Acceleration of Imperial Valley Earthquake

PEER NGA Strong Motion Database Record

Imperial Valley-02, 5/19/1940, El Centro Array #9, 180

Acceleration Time Series in units of g.

No of Points = 5372, Difference of Time = .0100 Sec

Scale Factor = I_g/R

Here, g = 32.2 ft/sec^2 = 386.4 in/sec^2

Scale Factor of Nonlinear Analysis , I_g/R_x = (1x386.4)/1=386.4

Linear Time History Analysis, I_g/R_x = 1×386.4/5 = 77.28

2.8.2 Scale Factor for Dhaka Region

Peak Ground Acceleration/ Maximum Considered Earthquake of Dhaka = 0.2g

Peak Ground Acceleration of Ground Motion = 0.2605g;

To scale down acceleration to $0.2g$ (Peak Ground Acceleration of Dhaka); let equalizing factor is A.

$0.2605g \times A = 0.2g$

So, $A = 0.768$

So, in Nonlinear Time History Analysis, Scale Factor = (386.4) (0.768) = 296.66.

In Linear Time History Scale Factor for 0.2g = I_g/R = 296.66/5 = 59.3

2.8.3 Scale Factor for Sylhet Region

Peak Ground Acceleration of Sylhet Region=0.36g

$0.2605g \times A = 0.36g$

So, $A = 1.382$

Scale Factor = 386.4×1.382 = 533.9

In Linear Time History Scale Factor for 0.36g = I_g/R=533.9/5=106.78

Result and Discussion

2.9 Introduction

This chapter highlights the analysis and results of 24 different finite element models prescribed in chapter 3. The result includes structural response from equivalent static analysis and nonlinear time history analysis. Results from equivalent static analysis for 24 frames are presented. Nonlinear behavior of 24 frames are discussed including computaion of deflection amplification factor and drift ratio. There are three different geometries/building configurations that includes 10m×10m, 15m×15m and 20m×20m floor area. Deflection Amplification Factor (C_d) and Drift Ratios are determined for all three geometric configuration of finite element.

2.10 Finite Element Analysis Result for 10m×10m Frame Structures

2.10.1 C_d and Drift Ratio of Two Bay Two Story Frame (2-2)

The selected beam of 235.2mm×416.6mm (9.26in×16.4in) and column of 235.2mm×235.2mm (9.26in×9.26in). Rebar percentage is 1%, 1.29% and 2.39% at bottom, first floor and second floor respectively. Demand Capacity Ratio is <1.

A. Deflection Amplication Factor (C_d)

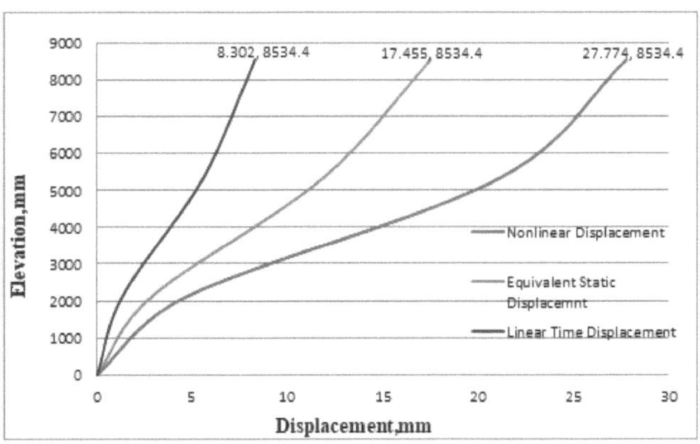

Figure 4.1. Maximum Story Displacement of 2 Bay 2 Story Frame

C_d = 1.59 and 3.35 respectively with respect to equivalent static displacement and linear time displacement.

B. Story Drift

As per BNBC, drift limit is 0.02. Here static linear, linear and nonlinear drifts are less than specified limit.

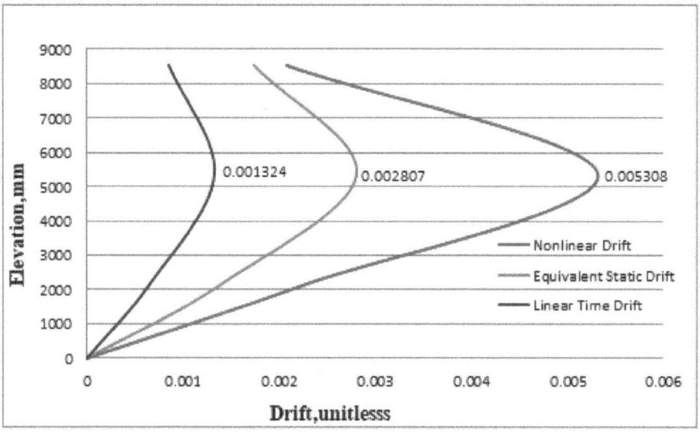

Figure 4.2. Maximum Story Drift of 2 Bay 2 Story Frame

C. Relationship between Displacement and Load

Two bay two story has maximum roof displacement of 27.8 mm where under load condition culumn has maximum displacement of 27.8 mm.

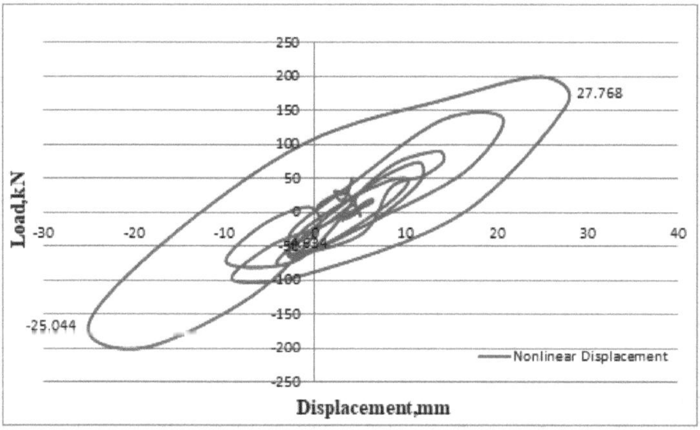

Figure 4.3. Relationship between Displacement and Load of 2 Bay 2 Story Frame

D. Hinge Result

The hinge result (Figure 4.4) shows that it is in plastic zone. Plastic rotation in negative direction.

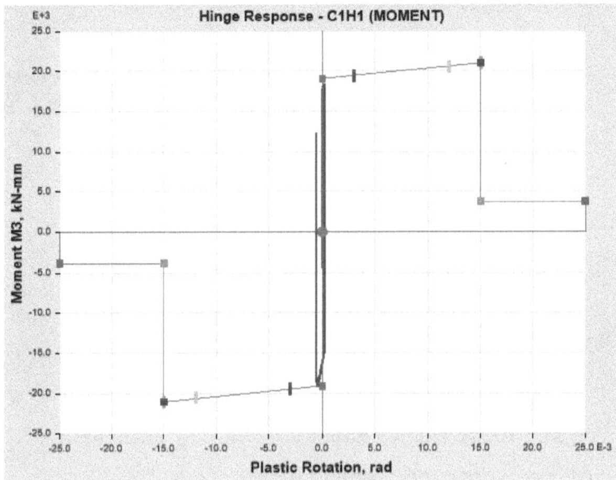

Figure 4.4. Hinge Result of 2 Bay 2 Story Frame

1.11 Evaluation of C_d for Sylhet (PGA = 0.36g)

2.11.1 Two Bay Two Story in Sylhet (2-2)

The selected beam of 235.2mm×416.6mm (9.26in×16.4in) and column of 235.2mm×235.2mm (9.26in×9.26in).

A. Deflection amplification factor (C_d)

The evaluated nonlinear displacement is 51.497 mm,the equivalent static displacement is 17.455 mm and linear time displacement is 14.944 mm. Therefore deflection amplification is found to be 2.95 and 3.45 with respect to equivalent static displacemnt and linear time displacement respectively.

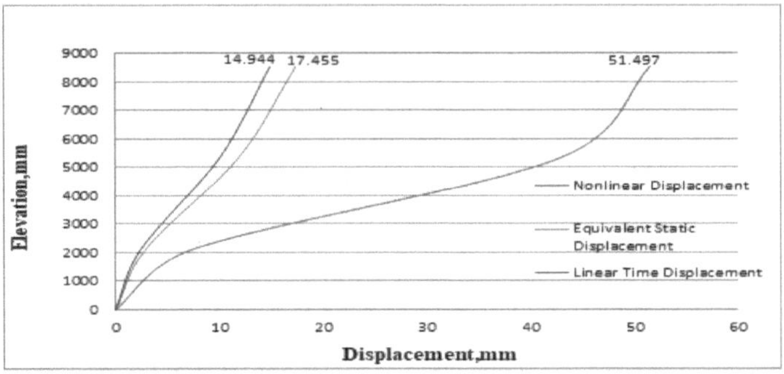

Figure 4.5. Relationship between Displacement and Load of 2 Bay 2 Story Frame

B. Hinge Result in Sylhet Region

The Load case is nonlineardynamic. The column is in ground floor. Hinge degree of freedom M3. Hinge relative distance is 0.1. The hinge has moved to plastic zone. Rotation is in positive and negative direction.

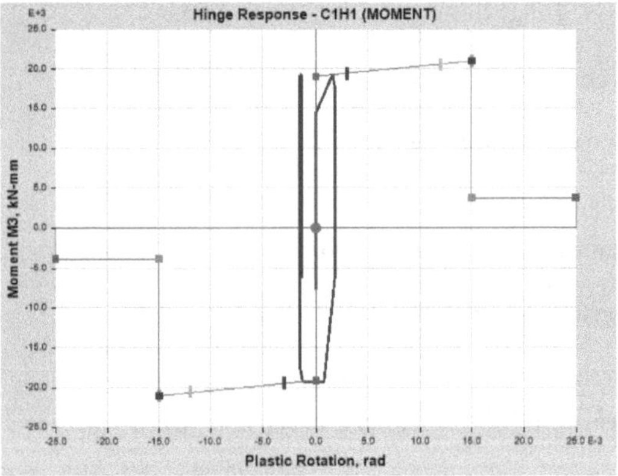

Figure 4.6. Hinge Result of 2 Bay 2 Story in Sylhet Region

2.12 Summary of C_d with respect to Equivalent Static Displacement and Storey Drift Characteristics (PGA = 0.3g)

Table 4.1. Summary of Inelastic Displacement Ratios and Storey Drift Characteristics

Frame	Analysis	Top Deflection (mm)	Story Drift	Inelastic Displacement Ratio
4 Bay 12 Story	Nonlinear Time History	110.13	0.008834	4.098
	Equivalent Static	26.873	0.000968	
4 Bay 10 Story	Nonlinear Time History	124.788	0.010601	3.03
	Equivalent Static	41.154	0.001804	
4 Bay 8 Story	Nonlinear Time History	85.494	0.00778	2.26
	Equivalent Static	37.772	0.002119	
4 Bay 6 Story	Nonlinear Time History	56.01	0.005339	1.84
	Equivalent Static	30.415	0.002287	
4 Bay 5 Story	Nonlinear Time History	49.443	0.00512	1.7
	Equivalent Static	29.112	0.002643	
4 Bay 4 Story	Equivalent Static	44.671	0.005402	1.79
	Nonlinear Time History	24.977	0.002801	
4 Bay 3 Story	Nonlinear Time History	42.229	0.006231	1.91
	Equivalent Static	22.096	0.003138	
4 Bay 2 Story	Nonlinear Time History	39.769	0.007563	2.02
	Equivalent Static	19.658	0.003652	
3 Bay 12 Story	Nonlinear Time History	103.313	0.006398	3.99

	Equivalent Static	25.888	0.000832	
3 Bay 10 Story	Nonlinear Time History	120.915	0.008206	2.98
	Equivalent Static	40.537	0.001638	
3 Bay 8 Story	Nonlinear Time History	82.916	0.006366	2.2
	Equivalent Static	37.717	0.001942	
3 Bay 6 Story	Nonlinear Time History	58.578	0.004579	1.81
	Equivalent Static	32.404	0.002192	
3 Bay 5 Story	Nonlinear Time History	49.935	0.004383	1.676
	Equivalent Static	29.795	0.002418	
3 Bay 4 Story	Nonlinear Time History	45.682	0.004768	1.736
	Equivalent Static	26.312	0.002632	
3 Bay 3 Story	Nonlinear Time History	41.99	0.005593	1.92
	Equivalent Static	21.87	0.002837	
3 Bay 2 Story	Nonlinear Time History	38.934	0.006694	2.07
	Equivalent Static	18.839	0.003179	
2 Bay 12 Story	Nonlinear Time History	173.152	0.006497	2.87
	Equivalent Static	60.388	0.001956	
2 Bay 10 Story	Nonlinear Time History	118.099	0.005895	1.87
	Equivalent Static	63.2	0.002554	
2 Bay 8 Story	Nonlinear Time History	75.478	0.00496	1.415
	Equivalent Static	53.332	0.002781	
2 Bay 6 Story	Nonlinear Time History	58.199	0.004473	1.17

	Equivalent Static	49.789	0.003521	
2 Bay 5 Story	Nonlinear Time History	74.167	0.00739	
	Equivalent Static	48.035	0.03758	
2 Bay 4 Story	Nonlinear Time History	53.85	0.006234	1.55
	Equivalent Static	34.72	0.003313	
2 Bay 3 Story	Nonlinear Time History	45.497	0.006903	1.78
	Equivalent Static	25.61	0.00309	
2 Bay 2 Story	Nonlinear Time History	40.808	0.008468	2.34
	Equivalent Static	17.455	0.002807	

2.13 Base Shear, Weight, Inelastic Displacement Ratios, and Storey Drift of Two Frames (PGA=0.3g)

Table 4.2. Base Shear, Weight, Inelastic Displacement Ratios, and Storey Drift of Two Frames

Frame	Analysis	Top Deflection (mm)	Story Drift	Base Shear, V (KN)	Total Weight, W (KN)	(V/W) %	Axial Load on Column (KN)	Column End, Rebar (%)	Column and Beam Size(in)	Inelastic Displacement Ratio
4 Bay 12 Story	Nonlinear Time History	110.13	0.008834	3728	50645	2.7	58174	0.82	21.75×21.75 and 12.30×24.75	4.098
	Equivalent Static	26.873	0.000968	1368						
4 Bay 10 Story	Nonlinear Time History	124.788	0.010601	2936	39310	3.7	45572	0.98	19.97×19.97 and 12.13×19	3.03

Conclusion

2.14 General

In this study the deflection amplification factor of Reinforced Concrete frames has been investigated and compared with values suggested in BNBC. For this, 24 frames were modelled in finite element software ETABS. These 24 frames were varied with respect to story height and bay configuration. Eight different story from 2 to 12 were considered in this work. Three types of bay arrangements were used, i.e., two by two bay, three by three bay and four by four bay.

The buildings were first designed using equivalent static analysis as per BNBC. From that beam and column sizes were determined. From this static analysis of 24 frames, maximum top deflection and storey drift were recorded. Afterwards, nonlinear dynamic analysis of these frames was conducted. For this, imperial valley earthquake data was used. This earthquake data was scaled to fit the BNBC recommendation for Dhaka zone, i.e., peak ground acceleration was seated to $0.2g$. For nonlinearity of members i.e., beams and columns, isotropic hysteresis model which is a built-in function of ETABS was used. From the nonlinear analysis, the top deflection, story drift and hinge results were recorded.

2.15 Conclusions

Following Conclusions may be drawn based on the study:

(i) C_d value was estimated from non-linear and linear top displacement found from the respective analysis. For 12,10 and 8 storied building had C_d value ranged from 4.57 to 4.23, 4.5 to 3.63 and 4.5 to 3.63 for 2,3 and 4-bay building respectively. BNBC specify a flat C_d value of 4.5 for buildings of all configuration. However, from this analysis work, it is evident that $C_d$value is very much dependent on building height, number of bays etc. For most of the cases analyzed in this work, C_d value was found to be much lower than BNBC specified valued. However, for 12 storied building, C_d was found to have value slightly more than 4.5 for 2 bay configuration.

(ii) C_d values suggested in this work may be useful for practicing engineers and researchers to estimate ultimate displacement of buildings due to severe earthquake from linear analysis of building for different regions of Bangladesh.

(iii) Cd value increases when Weight of building increases thereby probability of earthquake increases.

BISMILLAHIR RAHMANIR RAHIM

Chapter-3

Building Construction to Registration-A Critical Path with Solution Focusing Building Construction Act and Real Estate Law

3.1 Introduction

The population of Bangladesh is expected to rise significantly, and with it is being generated the need for yet new buildings. To cope with the situation and in an effort to improve conditions, recently, the regulations for buildings in Bangladesh have undergone a revision. The new regulations may help limit densification, based on area of plot, access road width and other physical parameters. To limit density, specific Floor Area Ratio (FAR) values have been assigned for various site conditions and functions. It is asserted that these changed regulations will be instrumental in improving the environmental situation. This paper attempts to investigate/analyze the extent of improvement that the new regulations make possible (with particular reference to residential development.)

In order to assess improvements, the situation in the immediate past is compared to the new one in terms of environmental aspects. For instance, setback needs in the past allowed buildings to occupy as much as 67% of site area, and there was no mandatory non-paved area for vegetation. The result was that almost the whole area could be hard-paved, and therefore there was a lack of any provision for water absorption, causing fast run-off and a consequent depletion in the water table. The new rules call for a specific percentage of mandatory green area for every site, addressing this concern. This in turn also affects heat absorption, temperature and humidity.

Building regulations are instrumental in shaping neighborhoods. In a city where land prices know almost no bounds[i], owners want utilisation of as much of the land as possible to maximise returns from the land as rent or for selling. This paper discusses the changes in regulations and the factors which have implications on the environment. It also analyses possibilities that the new regulations offer towards making residential buildings more energy efficient and environment responsive.

National Building Regulation of 1996

3.2. Building regulations were first introduced in Dhaka city[ii] through the Building Construction Act 1952 which basically was amended with slight changes the latest of which was the National Building Regulation of 1996. This was enforced until the Bangladesh National Building Code (BNBC) was made mandatory through national legislation in 2006. The discussion in this section concentrates on regulations in the immediate past with focus on residential construction and those codes that affected the environment.

Figure1: Setbacks guided layouts – monotonous development – little scope for ventilation and daylight

Set-backs guided layout (Figure-1). There was no mention of any Floor Area Ratio (FAR) limitations. Therefore there was no limit on the density of built form, and this could be as high as the height restriction of any area permitted. Studies[iii] during this situation show, with plot coverage of 67% average FAR was as high as 10.7. The concept of mandatory open space existed only as definitions of set-backs at the different sides of a building. The permitted set backs were as shown in

Table 1.

Table 1: ***REAR AND SIDE SETBACK REGULATIONS FOR RESIDENTIAL SITES IN DHAKA***

Site Area	Rear setback	Side Setback
Sites of 2 katha[1] or less	1.5m	Row houses possible
Sites of 3 katha or less	1.5m	1.25m
Sites of 4 katha or less	1.75m	1.25m
Sites of 5 katha or less	2.5m	1.25m
Sites greater than 5 kathas	3.0m	1.25m

[1] *Katha is a Bengali unit of land measure approximately equal to 720 sft (67sqm)*

The general tendency was to build on all permissible land, and these setbacks resulted in narrow strips of land on the different sides of each building[iv], unusable by the occupants (

Figure1). As the plots are mainly arranged in a grid iron pattern, the setbacks on back to back plots caused narrow strips of ventilation paths to be set up, resulting in funnelling – this creates differential velocity of moving air hampering the entry of air to interior spaces, where it could have been beneficial to thermal comfort.

6. It was also not mandatory to keep green areas on site.

7. Therefore, it was often common to pave these open areas. This resulted in increasing the albedo of the open areas, causing them to overheat when exposed to solar radiation.

8. Another effect was that they decreased the absorption of water – this resulted in faster run-off of rain water and prevented recharge of the water table. These effects caused the water table to fall considerably, which has in turn made Dhaka face a water crisis, as its main source of supply water is through tube-well extraction.

9. It also resulted in the creation of flash floods even with short spells and scant amounts of rain.

10. Another of the drawbacks of the regulations was that it had no mechanism to check construction and verify that it had followed the authorised plan was not adequate. This encouraged unlawful construction and infringement of regulations[v]. Buildings were often found to encroach on the mandatory open spaces, not abiding to the minimum setback limits set by the law – thus denying neighbours their air and lighting rights.

3.3 The Bangladesh National Building Code

However, the environmental effects could not be denied, as discussed in the previous section and the Bangladesh National Building Code was made legal and binding in 2006, over-riding all previous building regulations. The new legislation named Dhaka Metropolitan Building Construction Act, has been subsequently amended in 2007 and

2008 and is presently legally binding for all building activity in and around Dhaka, and is referred to as the Dhaka Metropolitan Building Construction Act, 2008. Key features of this legislation, which was first put down as the Bangladesh National Building Code, are discussed in the next section.

3.4 The Dhaka Metropolitan Building Construction Act, 2008

11. One of the key features of the new Dhaka Metropolitan Building Construction Act was to introduce the concept of floor area ratio or FAR – by which the density of settlements could be brought under control. To limit density, specific Floor Area Ratio (FAR) values have been assigned for various site conditions and functions. It is asserted that these changed regulations will be instrumental in improving the environmental situation of built areas and allow greater scope for passive design approaches aimed towards lowering energy consumption.

12. Another of the important changes that the new regulations brought about was a focus on providing for the safety of the occupants regarding fire hazards, strictly imposing regulations for the means of escape from a building and for access of fire trucks to all sides of the building on site.

13. Without the limitations of FAR, height restrictions imposed by the Civil Aviation Authorities, for safety related to flight paths, was the only control of density of built forms in many areas of Dhaka. Other than this, the RAJUK (*Rajdhani Unnayan Kartripakhya)* or Capitol Development Authority also stipulated certain height limitations for specific planned residential areas, like *Dhanmondi* R/A, where construction was limited to six storeys. The bar was removed on a few of the large circulation spines of the area, but not for plots bordering internal roads. Such restrictions were not imposed on informal residential areas, such as the areas neighbouring *Dhanmondi*, like *Sukrabad* or *Sobhanbag*. Thus in those areas, as long as the setbacks were maintained, buildings could go much higher. Similar was the case in the other formal residential areas of *Gulshan* and *Banani*. Densiites of these areas could therefore theoretically be quite high. Table 2 shows a comparison of density differences, among other key features affecting the environment, that the 2008 act allows in comparison with the 1996 regulations.

Perhaps a significant improvement over the 1996 regulations is provided in the monitoring aspect of the new Act. It is stipulated that architects will have to provide signed 'as-built' drawings which can be verified after construction of any project. This will certainly go a long way in stemming unlawful construction – as these drawings will be legally binding.

An example of the possibilities stemming from the consideration of FAR and other features of the new Act is given below.

3.5 Buildable area:

13. For an example site of 12mX18m dimensions, the area is 216 m2, under the new rules, permissible FAR is 3.5. Leaving area from three sides the allowable buildable area is 150 m2. It comes 69.44% of the total area.But maximum ground coverage is 62.5%.

3.6 <u>Calculation of Floor Numbers</u>

7. The Land size(12mX18m) is as shown below:

12m

a. Land Area=12mX18m=216 m2
b. Clear Area=10mX15m(After deducting clear distance
c. Ground Coverage=150/216=69.4%(From Chart Maximum Ground Coverage62.5%)
d. FAR for 216 m2 area=3.5
e. FAR=Total floor Area/Plot Area
   So,Total floor Area=FARXPlot Area=3.5X216
   So Total Floor area=756 m2
   Total Land 216 m2
   Area of One Floor 216X62.5%=135 m2
   So Floor Number =756/135=5.6=G+6 Floor

3.7 Benefits

The case demonstrates that the amount of free ground area is considerably increased by the new rule. Therefore using the new FAR, this site can get 3122 sft or 52.77% as open space. This is a considerable improvement in terms of the ratio of open:builtup space of previous rules. The larger open space consolidates the green into a more usable and defined shape.

Another improvement in the new rules is the variation that it encourages in the texture of the area (

Figure 2). Any number of possibilities are presented, compared to the monotonous setting of the past. Within the limits of the allowable buildable area, different clients will have different preferences – some will require more area per floor, some many opt for more floors and hence apartments for renting or selling and so on. So instead of the monotonous development pattern of the past, the new regulations allow and encourage variations. For this particular site the options were: 3143sft 8 storey buildings beside 2794 sft nine storey buildings alongside 3500 sft 7 storey buildings.

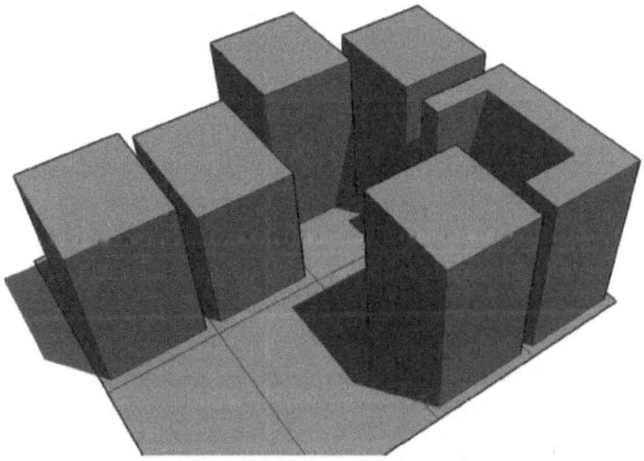

Figure 2: Variations possible under 2008 rules – ground coverage yields usable open spaces

Moreover, the ground area can be further increased by varying floor coverage, having less area on the lower floors and adding the saved area to the higher floors, by cantilevering, thus keeping even more of the ground free – provided the set-backs are not violated and one keeps within the permitted FAR and height limits.

An additional stipulation within the new rules states that half of this mandatory open space must be non-paved –leaving 20% green for recharging the ground water table. In general, with more open space to play with, some may have them in the front, some more privately at the back, while others may create introverted spaces.

If the neighbouring walls are removed, the quality of space further improves. There is thus variation in height, in building-to-building spacing as well as on the street frontage in terms of the building massing and layout (Figure-2) – a thing that was getting increasingly rare in the way residential areas were developing under the previous regulations. The main principle is that since the allowable buildable area is fixed, there is no clear advantage in building on every available open area, as then the building will merely be lower. The landowner soon realises that the advantages of larger open spaces in terms of the environment created are far greater than occupying the entire ground area.

3.8 Environmental impacts and Energy Efficiency

The main impacts of the new Building Construction Act of 2008, on the physical, spatial and visual quality of future residential areas, have been highlighted in the above section. This section will discuss effects on the environment and how these will serve to improve performance of the built environment towards achieving energy efficiency. For the example site, Table2 highlights the key features of the Act with comments representing the changed situation.

3.9 Density of Built Form

One of the biggest changes that this new Act brings about is a change in the density of residential areas. The control is brought about through the Floor Area Ratio (FAR) stipulations, which are strictly adhered to during the issuance of planning permission.

Density has direct impact[vi] on the solar radiation exposure. More open spaces in general increases the exposure – but with prior planning the exposed area can be oriented to face directions which will do the least damage to the resulting thermal condition.

Density also impacts the wind flow potential of areas through the presence or absence of obstacles to the regional flow pattern. Excessive ground coverage is also responsible for low water absorption, lowering humidity and raising rain water runoffs. Table2: Comparison between 1996 and 2008 Building Construction Regulations on the environment of Residential Areas

Features	Old regulations	New regulations	Comments on new regulations
Density of built form	High – limitless for areas without height limitations	Restricted according to FAR restrictions	Improves ventilation potential of site' Allows more consolidated open spaces
Open area/covered area ratio	33%	40%	This is referred to as mandatory open space – allows more consolidated open spaces
Paved area/unpaved area	Unspecified – hardly any unpaved	50% of mandatory open space	Specifying unpaved areas allows rainwater to recharge the ground water table Prevents rainwater runoff and flooding Lowers air temperature and thermal comfort ventilation possibilities
FAR	Limitless – as high as 10.7	Limited to fixed values for function and site access – 4.25 in example	Controls density and improves potential for thermal comfort ventilation Consolidates open spaces into usable spaces
Building-building distance	Based totally on setbacks	Variable and high on certain sides – as long as min setbacks are not violated	Improves thermal comfort ventilation potential Creates variation in built form Allows more consolidated open spaces
Variation in development/ morphology	Very low	High potential	Because FAR and ground coverage is regulated there is tremendous scope for different heights and spreads of buildings, resulting in diverse settlements.

3.10 Ratio of open area to covered area

The ratio of open area to covered area will determine the heat absorption of solar radiation and the consequent temperature regime of the site, especially as a significant portion of the open area needs to be unpaved. The more the green area the lower is the temperature escalation due to absorption, much of the heat in such cases is used in the biological processes of the vegetation.

Another of the benefits of green areas is its ability to recharge the water table by allowing absorption of rain water. This is extremely important in Dhaka where the water table has been reported as falling steadily due to extraction and low absorption.

3.11 Mandatory unpaved area

Since there was no mandatory unpaved area in the previous rules, there was no need to provide for vegetation. Highly paved neighbourhoods therefore developed increasing heat absorption and reducing rainwater absorption. This had the two-pronged effect of raising the temperature, and making flash floods more likely due to rainwater runoffs. Under the new rules, the temperature and flash floods can be kept under control to a greater extent.

3.12 Building to building distance

The greater the building to building distance, the larger is the possibility of airflow through settlements. The new rules allow wider spacing, as ground coverage can be reduced.

Varying building to building spacing, likewise, is more likely to permit variations in wind movement which can permeate the living spaces and create thermal comfort ventilation possibilities.

3.13 Balcony/ verandah concessions

Another of the new rules deals with getting concessions for balconies (which are defined as overhanging spaces with railings but no walls on three sides) – as opposed to verandahs (which are open on only one side). Such stipulations encourage spaces with greater indoor outdoor relationships than previously, with greater possibilities of indoor ventilation.

3.14 Scope for further improvement

The new Act of 2008 is now fully operational. There however remain a few issues where amendments may further improve environmental conditions. It is true that density is more controllable under the new Act. But to improve the usability of the open spaces which are now mandatorily kept unpaved, it is important to make some more changes, some of which are highlighted below.

Permission for semi-detached housing – by allowing plots to share party walls – such as is prevalent in many countries of the World.

Staggering plot layouts in new townships – this will improve the possibility of ventilation by utilising air funnels.

Passive design strategies should be quantified and the Act should build in trade-offs with proportionate increased maximum buildable area if it can be demonstrated that passive solutions are lowering dependence on conventional energy sources.

Environmental regulations for energy efficient options – building design should be encouraged to make use of renewable energy to supplement conventional energy usage. As this may involve additional initial installation costs, incentives from the Government may be needed as bonus. Some mandatory percentage of the buildings total energy consumption should be from renewable sources. As solar energy is an option, building integrated photovoltaics (BIPV) can be actively used.

Green measures using available resources should also be stipulated for receiving trade-offs. Daylight use can be actively demonstrated through simulations to demonstrate reduced dependence on conventional energy during the day for lighting.

Green measures like rainwater harvesting should also be encouraged – if dependence on the city's water supply can be lowered, it improves liveability of the city, reduces dependence on energy for extraction, as well as prevents depletion of the water table.

Green measures should also be taken to use and reuse resources within the building – like using grey water for gardening or cleaning

Building materials with low embedded energy should be identified, grouped according to manufacturing energy as well as transportation energy, so that points can be gained for greener alternatives.

Energy efficient appliances, like motion sensing lights, water taps, etc should be encouraged as fixtures to reduce wastage within the building.

The codes should address these issues so that the resulting building activities can encompass greener options. The code should devise a system by which earned 'greenness' index can be developed which can be exchanged for tax or floor area benefits.

and water conservation

Conclusions – early to predict how the situation will change – but strict adherence to the rules as architect needs to vouch for the built product – more is expected in terms of shouldering responsibility – as the implementers will be required to validate their work after building – therefore more accountability – violations are likely to fall – overall improvement but more research is needed

Cities have historically been seen as the antithesis of nature, intrinsically unnatural and destructive. DA08 Issue Summer 2008 Velux. P17 Beatley, T. Newman. P.

Nature must become a central paart of future urban design programmes.

Ciites must be deisgned for resilience – even if global emissions of greenhouse gases are substantially reduced, the effectrs of climate change will manifest themselves and be specially severe for cities – impacts range from heatwaves and droughts to extreme storms and rise in sea level.

3.15 Building Construction Act- 2008

2. Taking into cognigence of Building Construction Act,1952 (Section 18) the Dhaka City Building Construction Act 2008 is introduced.

a. Chapter 1. It mainly covers various definitions related to building construction

b. Chapter 2. It mainly covers Procedure for approval of drawings, Building Permit and Occupancy Certificate

c. Chapter 3. It mainly covers various Committee.

d. Chapter 4. This Chapter covers list of technical persons and Categories

e. Chapter 5. This Chapter Covers Building Construction Rules

f. Chapter 6. Health and Security

g. Chapter 7. Conservation, Preservation, Construction in Hilly, Forest and Water logged Area

3.16 Building Construction Rules

3. Mandatory open space, Floor Area Ratio, Maximum Ground Coverage and set back from boundary are main focus of building Construction Rule.

a. Floor Area Ratio
b. Maximum Ground Coverage
c. Set Back from Boundary

3.17 Floor Area Ratio

4. $F \quad A \quad R = \frac{\text{floor area}}{\text{lot area}}$

In practice, this ratio is constant for a zone. A floor area ratio of 1.0 means that floor area may equal lot area. FAR 5.0 means that the floor area may be up to five times as large as the lot area; and FAR 0.5 that it may be no more than half the lot area.

Certificate by Author Definition of Floor area ratio

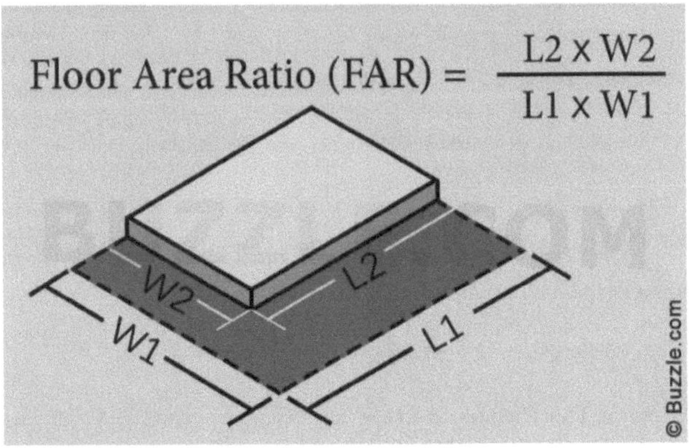

● Floor-Area Ratio (FAR)

$$FAR(\%) = \frac{\text{total floor area}(B+C)}{\text{site area}(A)} \times 100$$

● Building Coverage Ratio (BCR)

$$BCR(\%) = \frac{\text{building area}(B)}{\text{site area}(A)} \times 100$$

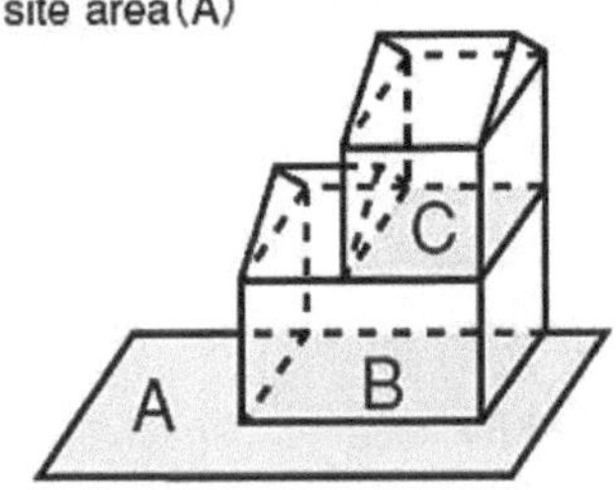

3.18 Calculation OF FAR

5. Formula for calculating FAR is quite Simple. "Total covered area of all floors divided by the plot area." **Floor area ratio = (total covered area on all floors of all buildings on a certain plot, Gross Floor Area) / (area of the plot).**

6. Plot Area= 60 feetX100 Feet= 6000 Square feet

Floor Area= 2100 Square Feet

Floor area rato= 2100 Square feet/6000 Square feet

So FAR=.35

3.18.1 <u>Maximum Ground coverage</u>

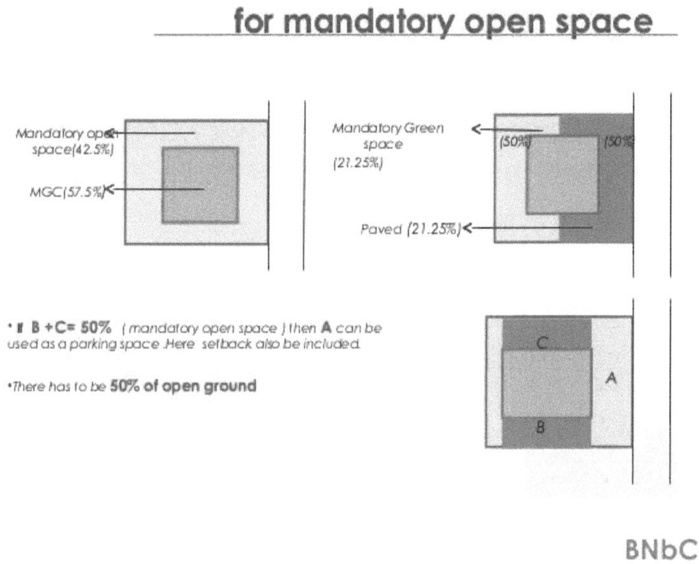

3.19. Building Height

After determining the allowable buildable area, the Architect can then design the building to any height dividing the allowable 25143 sft by the permissible ground coverage of 3500 sft – allowing 7.1 floors in this case. But any combination that allows higher buildings with lower ground coverage is also permitted within the rules. The only limiting factor in such cases is the Civil Aviation regulations that deal with the flight path funnels that limit the building heights in various zones around the International Airport in Dhaka.

This latter limitation for the example case was 150 ft above sea level, technically allowing a building of approximately 15 stories. The ground coverage in that case would be only 1676 sft. However considerations of appropriate size of residential apartments to permit renting or selling also sway the ultimate decision of height.

Among other considerations within the Code are the more stringent provisions needed for buildings of 10 storeys and above. These include provision of two stairs as fire escapes which has to be provided within the permitted buildable area. Owners therefore tend to favour limiting construction to within 9 storeys.

The combination of the considerations above led in the example case to a decision to build to 9 storeys, which led to ground coverage of 2794 sft. Moreover, there are some additional concessions that the owner gets for balconies and plantation areas.

(FAR) (MGC) x [Type: A (A1-A5) : আবাসিক বাড়ী ও হোটেল]

Measarment of Plot		Class of Building (A1-A4) Residential House			
Sqm	**Katha**	**Wide of road (Meter)**	**FAR**	**MGC (%)**	
134 sqm and below	02 Katha and below	6.0	3.15	67.5	
134 to 201 sqm	02-03 Katha	6.0	3.35	65.0	
201 to 268 sqm	03-04 Katha	6.0	3.50	62.5	
268 to 335 sqm	04-05 Katha	6.0	3.50	62.5	

3.20 Calculation of Floor Numbers

7. The Land size(12mX18m) is as shown below:

f. Land Area=12mX18m=216 m2
g. Clear Area=10mX15m(After deducting clear distance
h. Ground Coverage=150/216=69.4%(From Chart Maximum Ground Coverage62.5%)
i. FAR for 216 m2 area=3.5
j. FAR=Total floor Area/Plot Area

   So,Total floor Area=FARXPlot Area=3.5X216

   So Total Floor area=756 m2

   Total Land 216 m2

   Area of One Floor 216X62.5%=135 m2

   So Floor Number =756/135=5.6=G+6 Floor

3.21 Construction Rule/ Illegal Construction

1. Definition in Gazette. "Approved drawing" means drawings of buildings or structures as per Law.

2. Construction Can be Done in Ground Floor. A guard room having area of 60 square feet can be constructed in front of plot with attached toilet, not more than 8 feet height.

3. Construction Can be Done in Roof. A water reservoir, stair room(maximum 7 feet) can be only constructed in roof and a machine room(Height 13 feet) can be only constructed for use of lift in the building.

4. Cantilever Veranda.

5. Construction beyond the Drawing.

3.22 Actions Against Illegal Construction

3.23 Real Estate Laws in Bangladesh For Buying/Selling

2.As you already know there are specific terms and conditions regarding buying and selling of any property and both parties must abide by them. So, let's not bore with the things you know and focus on things that you may not be aware of generally. As per the contract, the buyer or seller cannot change the quoted price. However, the amount can be adjusted later if the parties mutually agree on using better materials for the property. Next, the developer must handover the property, registration and the principal deed in no more than three months. Also, during transfer, if the property has any deviation regarding the size or space, the price must be adjusted within three months of the handover. Therefore, next time when you buy a property, keep these things in mind.

3.24 Regarding The Property Contract

3. For any transaction, there has to be a legal property contract. Again, this is also a common thing. So here are the important laws that you should know. Firstly, as a landlord, you will get a particular portion of the apartment building for your ownership. However, if you want to sell off the apartments of your share, you will have to execute a power of attorney and give the right to the developer to sell your part of the property. Then, as an owner of your new apartment, you are entitled to enjoy one year of the maintenance facility by the developer. Moreover, if there are any defects within two years after the handover, don't worry. The developer must repair the damage from their pocket and cannot ask for your contribution as the owner.

3.25 Drawings to be counter Signed by Developer

4. Except the approved drawing other associated architectural design, the structural design, Other building Services design and deeds to be Counter signed by the developer(Real Estate Development and Management Law 2010, Chapter Two,Para 6, Subpara 5).

3.26 Real Estate Construction by Landowner

5. An individual or few people jointly can build Real Estate on his own land. The Real Estate Development and Management Law 2010 would be applicable for him/them also.

3.27 Registration to Buyer (Real Estate Law)

6. Developer has to handover occupation and give registration to buyer within three months of paying total dues.

7. Developer can provide deed to buyer once power of attorney is given to developer by the landowner.

8. The landowner should give reqistration/provide deed within 15 days after getting written request from developer if developer is not given power of attorney.

9. Developer can provide deed to buyer if land owner does not provide the deed once requested by the developer.

3.28 <u>Preparation and Signing of Drawings</u>

10. All drawings submitted for approval shall be prepared and signed by registered professionals as specified in Table 2.3.4 of Bangladesh National Building Code 2015(Final Draft),which shall be considered as equivalent to certifying that the drawing on which the signature appears conforms to all the requirement of the Bangladesh National Building Code. Registered Professionals shall put his or her signature with date on the title box of the drawing along with his name, address ,professional society membership number, registration number and any other information required by the concerned Building Official.

11. The drawing shall also contain the signature,name and address of the owner.

12. Subject to the classification and use of buildings,all drawings for approval and execution shall be prepared and signed by the registered professionals as per building category specified in Tables 2.3.3 and 2.3.4 corresponding to relevant work.

---

i **The FINANCIAL EXPRESS** (2008) Dhaka sees unprecedented rise in land price. daily newspaper – internet version. Home page. Dhaka. 17 May 2008. www.thefinancialexpress-bd.com accessed on 28 Jan 2010

ii **CHARKRABORTY, P.** (2008) The Building Construction Act and Rules. Law Book Pavillion. Dhaka.

iii **KAMRUZZAMAN, M.** (2007) Potential of High Density Walk-up Apartments in Dhaka City. Research Report to the Architectural Institute of Japan (AIJ)

iv **AHMED, Z.N.** (1994) Assessment of Residential Sites in Dhaka with respect to Solar Radiation Gains, unpublished Ph. D. Thesis; De Montfort University; UK; ch2

v **MAHMUD, M.A** (2007) Corruption in Plan Permission Process in RAJUK: A Study of Violations and Proposals. Transparency International, Bangladesh. Dhaka. Aug 2007 p6-7

vi **Z.N. AHMED; G.S. ROY**. (2007) <u>Ventilation Potential: Examining the Effects of Growing Densification in the Tropics</u>; proceedings International Seminar on Well Being Indoors: Clima 2007; Finland, 10-14 June 2007

FSC
www.fsc.org
MIX
Papier | Fördert
gute Waldnutzung
FSC® C083411

Kolibri 360 GmbH
Ferdinand-Jühlke-Straße 7
99095 Erfurt, Deutschland
produktsicherheit@kolibri360.de

Druck:
CPI Druckdienstleistungen GmbH
im Auftrag der
Kolibri 360 GmbH
Ferdinand-Jühlke-Str. 7
99095 Erfurt